BIG WORDS FOR LITTLE PEOPLE

Helen Mortimer & Cristina Trapanese

Nature

OXFORD
UNIVERSITY PRESS

Living things

All around us there are living things. Nature is everywhere. We can see it, hear it, smell it and feel it!

. . . to this big tree.

Plants

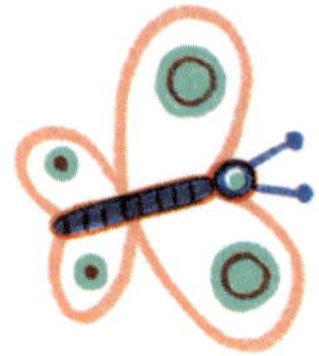

Flowers, grasses, ferns, moss, bushes and trees are all plants.

BZZZ! A toadstool isn't a plant. It's a fungus!

Animals

Animals come in all shapes and sizes.
Bugs, birds and pets love gardens.

BZZZ! I'm an insect.

Life cycles

Plants and animals have different life cycles. Here are just two.

frog
frogspawn
tadpole
froglet
frog

Growing

Plants need sunlight and water to grow. Animals need food and water.

We all need love and care!

Senses

Animals use their senses to find food and stay safe.

Cats have excellent hearing.

Habitats

Nests, webs and ponds are homes for different kinds of animals.

BZZZ! Some animals live underground, like worms, but I live in a nest.

Behaviour

Some plants creep up walls.

Some animals are awake at night – they are nocturnal.

We all behave in different ways and have different habits.

Rocks and soil

Our planet is made from rock. It's what is underneath our feet.

Soil is made from tiny bits of rock mixed with rotting plants.

We love mud!

Caring for nature

There are lots of ways we can look after nature, such as by planting seeds, making bug hotels and putting up feeders.

Birds need baths too!

Nature in danger!

Some plants and animals are very rare. We need to protect the places where they live.

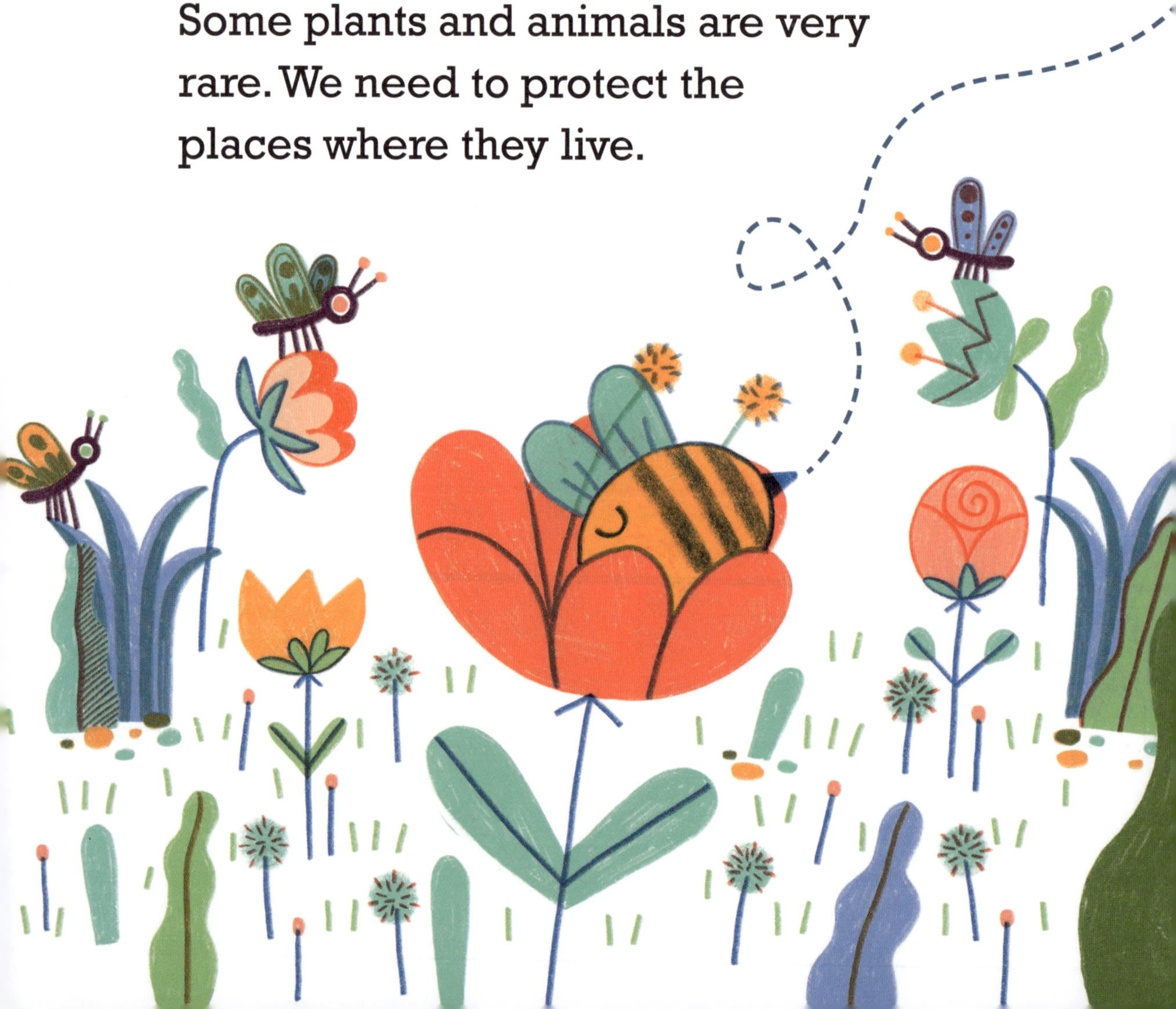

BZZZ! This is my happy place!
This is a rare butterfly!
We should never pick wild flowers!

All about nature

What do you do to look after nature?
Can you name five different trees?
Which animal do you like best?

Ten ideas for getting the most from this book

1 Find somewhere cosy and comfortable to share this book. It's important not to feel rushed.

2 Remember that science isn't something that happens only in a classroom – there are opportunities all around us to talk about science concepts and use scientific language.

3 Ask each other all kinds of questions about nature as you share the book.

4 The children in this book are in an outdoors setting. Remember to look for animals, plants, habitats and features of the landscape when you're out and about.

5 Collect leaves, feathers, and seeds from the ground and make some nature-inspired collage art.

6 Try growing something from seed on a windowsill – cress seeds are a good option.

7 Listen to the birds when you are out for a walk – how many different bird songs can you recognize?

8 Talk about where the vegetables and fruits in your fridge come from and what they look like when they are still in the ground or growing on plants.

9 Notice how nature changes with the seasons.

10 Use local wildlife organizations or nature reserves to find out more about the natural world in your neighbourhood.

Glossary

fungus – a living thing, such as a mushroom, a toadstool or mould, which has no leaves or flowers and grows on plants or rotting material

insect – an animal with six legs and one or two pairs of wings

habitat – a plant's or animal's habitat is where it lives

sprout – the young shoot of a plant

rare – if things are rare, there are not many of them and so they are very special